Floor Collapse Claims Two Firefighters
Pittston, Pennsylvania

Investigated by: J. Gordon Routley

This is Report 073 of the Major Fires Investigation Project conducted by TriData Corporation under contract EMW-90-C-3338 to the United States Fire Administration, Federal Emergency Management Agency.

Department of Homeland Security
United States Fire Administration
National Fire Data Center

U.S. Fire Administration Fire Investigations Program

The U.S. Fire Administration (USFA) develops reports on selected major fires throughout the country. The fires usually involve multiple deaths or a large loss of property. But the primary criterion for deciding to do a report is whether it will result in significant "lessons learned." In some cases these lessons bring to light new knowledge about fire--the effect of building construction or contents, human behavior in fire, etc. In other cases, the lessons are not new but are serious enough to highlight once again, with yet another fire tragedy report. In some cases, special reports are developed to discuss events, drills, or new technologies which are of interest to the fire service.

The reports are sent to fire magazines and are distributed at National and Regional fire meetings. The International Association of Fire Chiefs assists the USFA in disseminating the findings throughout the fire service. On a continuing basis the reports are available on request from the USFA; announcements of their availability are published widely in fire journals and newsletters.

This body of work provides detailed information on the nature of the fire problem for policymakers who must decide on allocations of resources between fire and other pressing problems, and within the fire service to improve codes and code enforcement, training, public fire education, building technology, and other related areas.

The Fire Administration, which has no regulatory authority, sends an experienced fire investigator into a community after a major incident only after having conferred with the local fire authorities to insure that the assistance and presence of the USFA would be supportive and would in no way interfere with any review of the incident they are themselves conducting. The intent is not to arrive during the event or even immediately after, but rather after the dust settles, so that a complete and objective review of all the important aspects of the incident can be made. Local authorities review the USFA's report while it is in draft. The USFA investigator or team is available to local authorities should they wish to request technical assistance for their own investigation.

This report and its recommendations were developed by USFA staff and by TriData Corporation, Arlington, Virginia, its staff and consultants, who are under contract to assist the USFA in carrying out the Fire Reports Program.

The USFA greatly appreciates the cooperation received from the Pittston Fire Department, with special thanks to Chief Louis Calabrese, Assistant Chief James Rooney, and Assistant Chief Frank Roman for the information and assistance they provided.

For additional copies of this report write to the U.S. Fire Administration, 16825 South Seton Avenue, Emmitsburg, Maryland 21727. The report is available on the USFA Web site at http://www.usfa.dhs.gov/

U.S. Fire Administration
Mission Statement

As an entity of the Department of Homeland Security, the mission of the USFA is to reduce life and economic losses due to fire and related emergencies, through leadership, advocacy, coordination, and support. We serve the Nation independently, in coordination with other Federal agencies, and in partnership with fire protection and emergency service communities. With a commitment to excellence, we provide public education, training, technology, and data initiatives.

TABLE OF CONTENTS

Floor Collapse Claims Two Firefighters
Pittston, Pennsylvania

Local Contacts: Chief Louis Calabrese
Assistant Chief James Rooney
Assistant Chief Frank Roman
Pittston Fire Department
20 Kennedy Street
Pittston, PA 18640

Fire Marshal Sylvester Myers
Pennsylvania State Police
Chief Edward Doran
Pittston City Police Department

OVERVIEW

Two volunteer firefighters were killed in the early morning hours of Monday, March 15, 1993 in the town of Pittston, Pennsylvania. The two firefighters, who were members of separate departments, were operating as a team on a hoseline, attempting to locate a concealed fire, when a sudden and unanticipated floor collapse sent them crashing down into an inaccessible combustible concealed space. Even though both were wearing full protective clothing, using self-contained breathing apparatus, and operating with the protection of a handline, they were unable to escape from the building or find refuge from the rapidly advancing fire conditions. Rescue teams were unable to reach the victims due to difficult access and rapid fire spread throughout the fire building and interconnected structures.

The two firefighters who died were Captain John F. Lombardo of the Pittston Fire Department, age 26, a six year fire service veteran, and Assistant Foreman Leonard lnsalaco II of the West Pittston Fire Department, age 20, a two year fire service veteran.

SUMMARY OF KEY ISSUES

Issues	Comments
Situation	Fire in a concealed space below the ground floor level. Crews had difficulty locating the fire in the complicated structure.
Structural Collapse	Floor collapsed, dropping two firefighters into fire area, moments after flames were located.
Rescue Efforts	Rescue efforts were unsuccessful due to lack of direct access to fire area and rapid fire spread throughout structure and exposure.
Fire Control	Entire complex of interconnected structures became involved. Elevated master streams were used to confine and control fire.
Building Condition	Structures were more than 100 years old, with numerous renovations, changes of occupancy, interconnections, and previous major fire. No pre-fire plan information available.
Accountability	Identity and number of missing members in doubt due to lack of accountability system. Entry crew had PASS units, but no radio communications. Crews were assembled at the scene from personnel who responded.
Access to Fire Area	No access from occupancy above to fire area below. Only access was through vacant occupancy on lower level with entry from street at rear.
Communications	Radio system is inadequate for the needs of the fire department. Entry crews did not have portable radios to communicate with Incident Commander.
Pre-fire Plan	No pre-fire plan was available to assist the fire chief in directing operations. The complicated buildings presented unique problems that could not be visualized without a plan.

The purpose of this report is to provide educational information for the fire service, with the hope that future accidents of a similar nature may be avoided. It is not intended to find fault with the actions of any individual who was involved in the operations or to fix responsibility for the fire or the deaths that resulted.

The review of this incident will note several lessons learned as a result of this tragedy, many of which are similar to the observations from previous incidents. It is the intent of this report to provide an educational basis from which these lessons can be learned by the fire service, so that it can better prepare and equip itself for future missions.

There is an inherent level of risk that will always be present in the operations of a fire department at the scene of any emergency incident. Through training, education, and experience, fire service members can be better prepared to anticipate the outcome of all types of incidents and to react to the circumstances that they are presented with in each situation. The officer in command of a fire must be able to identify the risk factors that are present in a given incident and formulate a strategic plan that takes all of those risk factors into consideration. The fire service must also be prepared to react to unexpected situations and conditions.

The firefighters who died in this incident were trained and experienced and were operating in what they considered to be a normal situation with a normal approach to operational safety. The experience of this incident should be carefully considered by every firefighter and particularly by present and future incident commanders.

FIRE SERVICE ORGANIZATION

Pittston is a community of approximately 9,500 people located 10 miles south of Scranton, Pennsylvania on the east edge of the Susquehanna River. Fire protection is provided by the Pittston Fire Department, which is comprised of two separate volunteer companies, Eagle Hose Company No. 1 and Niagara Engine Company No. 2. Pittston City employs seven career personnel; the fire chief, two assistant chiefs, and four drivers. The fire chief has overall authority and responsibility for all operations and the Pittston City owns the apparatus.

The Department is directly supported by the city from property tax revenue. The volunteer firefighters receive no compensation and there is no operational distinction between the two companies at the scene of a fire; all members operate under the direction of the city fire chief and the two assistant chiefs. The single fire station belongs to one of the volunteer companies but is staffed by the career personnel. Even though the community is suffering from severe economic conditions, the volunteer organizations are reported to be very stable. The volunteer companies have invested several million dollars of revenue from the Pennsylvania Fireman's

Relief Fund which is used to purchase safety equipment and to provide for the health and welfare of the members and their families. This is exemplified by the recent purchase of 40 sets of new turnout gear and the use of state-of-the-art self-contained breathing apparatus and personal alarms (PASS devices).

The Pittston Fire Department operates two Class A engines and an elevated platform aerial device. A vehicle equipped with spill control material is shared by the police and fire departments and is housed at the combination city hall and police station, approximately two blocks from the fire station. Emergency Medical Services are provided by an independent volunteer rescue squad with its own station.

The volunteer fire companies have about 100 members of which an estimated 40 to 50 are active firefighters. The minimum career staffing is two personnel on duty at all times, with the workload shared evenly by all seven employees. Volunteer members are hired as part-time employees on an as needed basis to cover for absent career personnel or supplement the staffing due to unusual conditions that might exist.

The normal response to structure fires is to have the on-duty career personnel respond to the scene with the two pumpers, accompanied by any volunteers who happen to be at the station. Alarms are dispatched by the Pittston Police Department and all personnel are alerted via radio receivers. The volunteer and off duty career personnel normally respond directly to the scene to meet the apparatus. There are no predesignated crews and it is up to the officer in command to organize the arriving personnel into operational teams.

The neighboring volunteer departments in West Pittston, Exeter, Jenkins Township, and Duryea provide mutual aid to Pittston on request of the Incident Commander. The career departments in the cities of Scranton and Wilkes Barre, located north and south respectively, are also available on mutual aid. Additional assistance can be requested from other volunteer departments in Luzerne County.

During the 36 hour period prior to the fire a heavy snowfall, with strong winds and bitter cold, had paralyzed the community. Snow removal crews had begun to clear the streets, but most were limited to a single lane and many hydrants were buried in snow.

On the night of the fire the fire chief had exercised his authority to hire two volunteers as part-time employees to increase the crew at the station from two personnel to four. The Chief felt that due to the severe snow conditions the extra personnel would be needed to assist the drivers with their routine duties of hydrant spotting, layout, and possibly digging hydrants out of the snow banks caused by drifting and plowing.

The fire station is located approximately two blocks from the fire scene, adjacent to the main business district, where the fire building was located. All three major pieces of apparatus operated by the department are located at this station. There is no command post vehicle provided for the chief officers to utilize and they therefore are required to rely on their personal vehicle for response to the fireground. This shortage severely limits the resources that are available to the Incident Commander in his responsibility to manage this or any incident.

FIRE BUILDING

The fire occurred in a complex of buildings in the crowded central commercial district of Pittston, fronting on North Main Street. The complex was comprised of four original structures that had been interconnected over the years.[1] (The block plan, ground floor plan with dimensions, and a cross section diagram appear on the following pages.) At the front there were two single story storefront occupancies, an optical service establishment and a stationery store. At the rear there were entrances to a pizza parlor and a vacant office space, opening onto Crom Street, each with a story above. Due to the differences in elevation from Main Street to Crom Street, the street level facing Main Street coincided with the upper level at the rear of the buildings. The street level entrances at the rear were approximately level with the basement at the front of the buildings.

The four original structures appeared to be more than 100 years old and had been altered many times over the years. At the front the two original structures were three or possibly four stories in height, but the upper floors had been removed after a fire that is believed to have occurred in the 1950s. The buildings were essentially twins and appeared to have been constructed together. They had thick brick outer walls and a pair of back-to-back double course brick walls extending from front to rear, physically separating the structures into two separate buildings. Their narrow width appears to have been a detail necessitated by the construction method, since the floors and roof were supported by heavy wood joists spanning approximately 25 feet between the brick walls. The upper levels had included a public assembly occupancy that is reported to have extended through both buildings and there was evidence of several openings at the ground level that had existed at one time or another as the occupancy of the buildings changed.

Below the street level of these buildings was a basement level. The ground floors were wood supported by exposed 3 by 10 wood joists spanning the width between the brick walls under each occupancy. The clear height of this space was reported to be too short for normal occupancy, but it was used for storage. Under the vision center this space had been partially finished, but was unoccupied. Below the basement levels there were unfinished sub-cellars.

The rear occupancies were originally individual structures, separated from the front buildings by an alley. The rear buildings also may have been twins and the construction details were similar to the front buildings. These structures had street level entrances from Crom Street with a cellar below and a story above.

---

[1] No detailed plans could be located for the fire buildings, so all descriptions are taken from verbal descriptions and visual examination of the rubble.

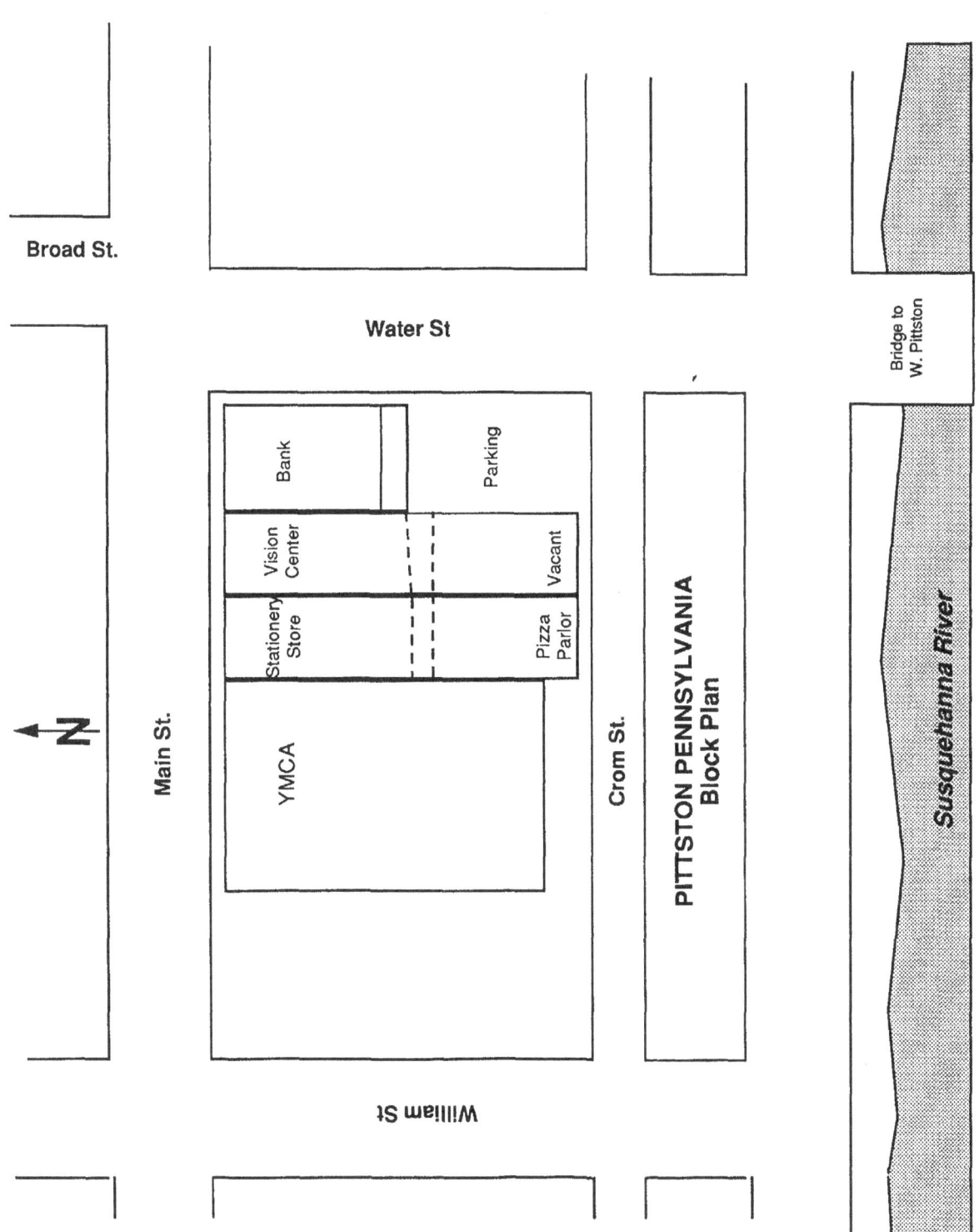

PITTSTON PENNSYLVANIA
Block Plan

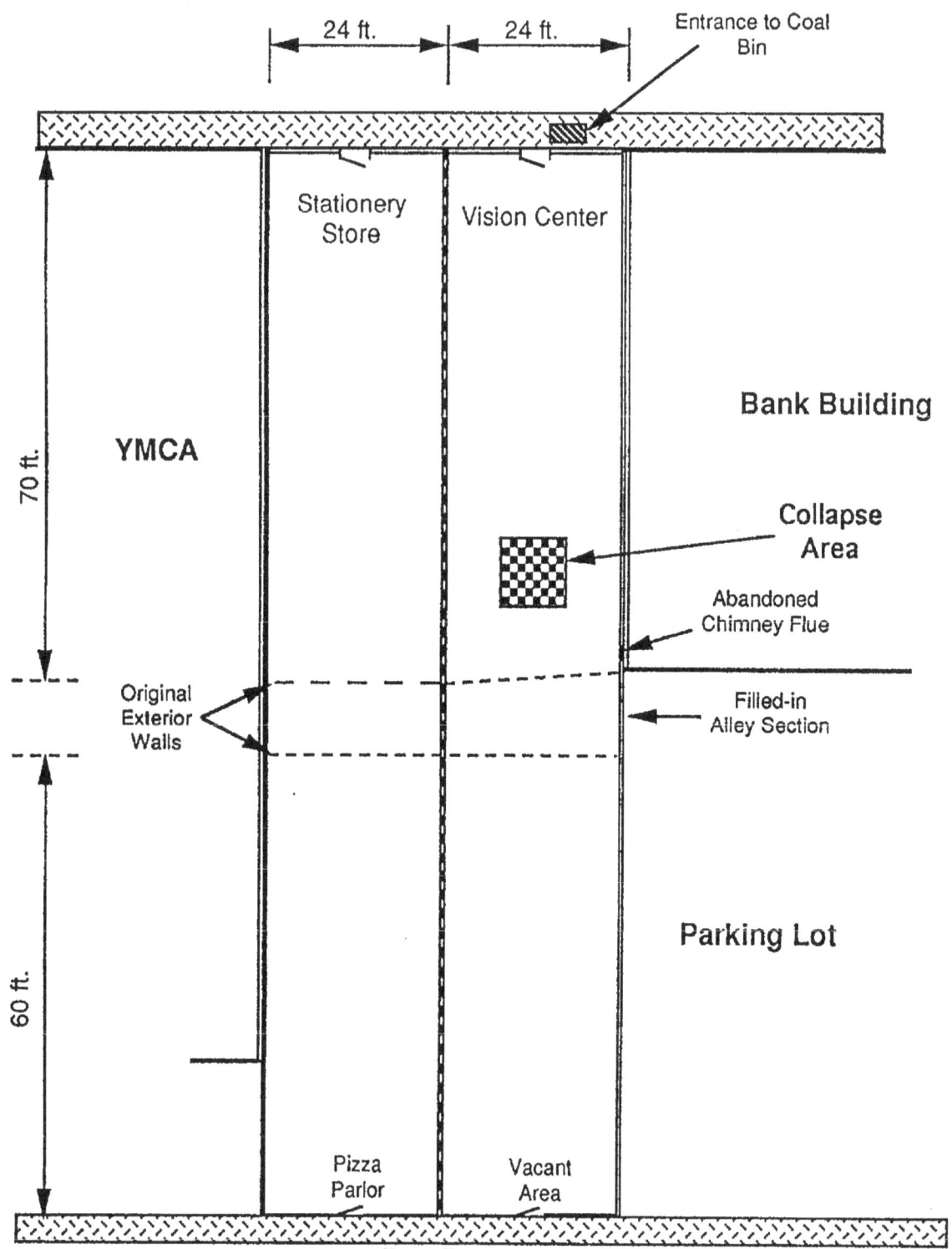

ESTIMATED DIMENSIONS

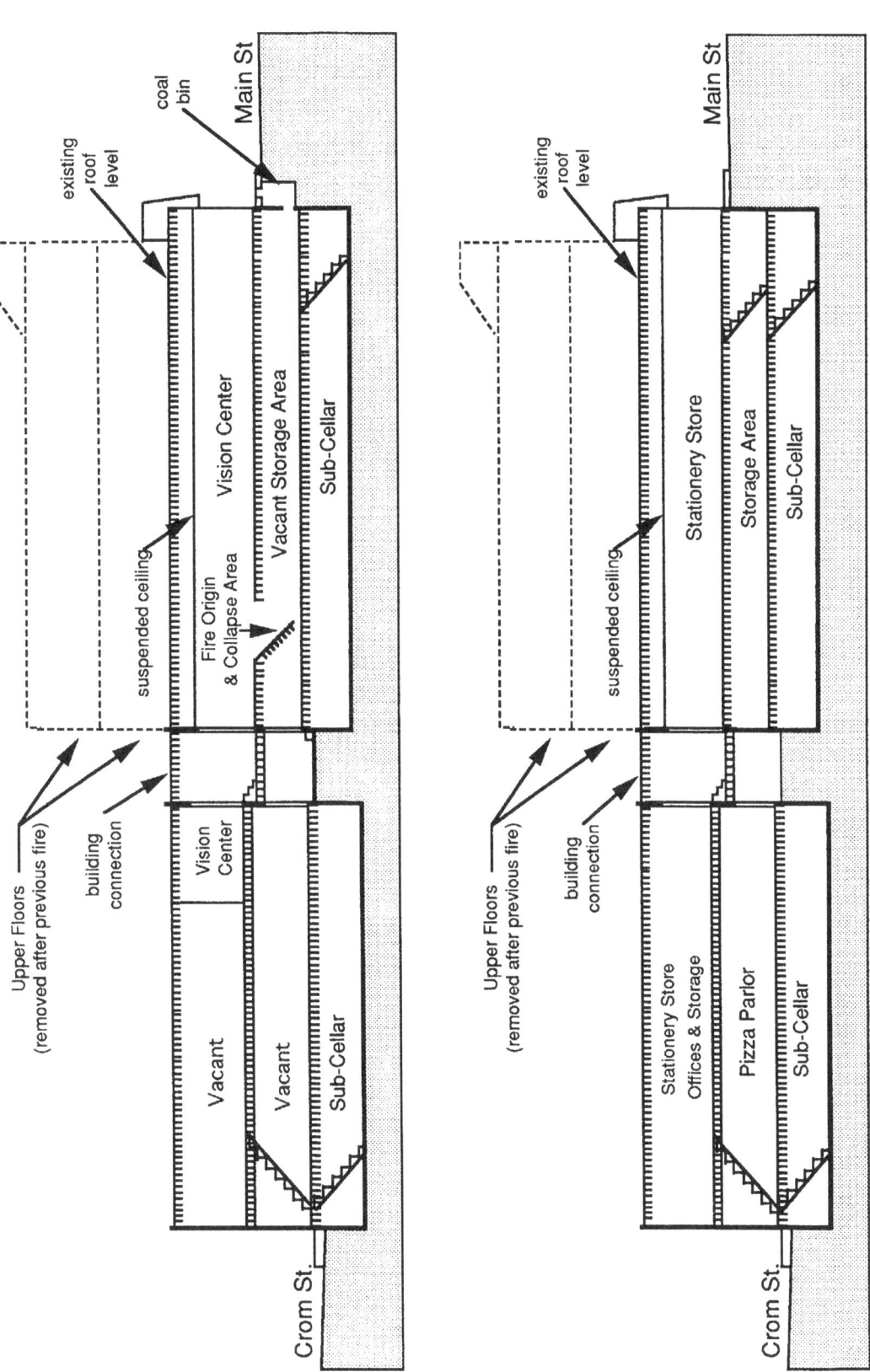

PITTSTON, PENNSYLVANIA
Cross Sections through Occupancies

At some point the alley between the two structures was closed and built over, linking the front and rear structures. The rear walls of the original buildings became interior walls and a new section was built, linking the street level at the front with the upper floor of the rear structure, and the street level at the rear with the basements of the front buildings. The cellars did not connect, as there was no cellar space where the alley had been.

It is believed that the interiors of the two buildings were altered numerous times over the years as the occupancies changed. The stationery store extended back into the upper level of the structure behind it and the vision center occupied part of the space on the upper floor of its rear building. A pizza parlor was located in the lower rear occupancy, under the rear of the stationery store, and was operated by a relative of the stationery store owner. The two levels were linked by a stairway that connected the rear of the stationery store with the pizza parlor. The stationery store also had an access stair leading to the storage area under the main part of the store.

The rest rooms for the pizza parlor were new construction, built into the space that had been the alley connection between the two buildings, suggesting that there had been a large opening between the lower occupancies at some time in the past. There was another large opening between the rear structures at the upper level that had been blocked by wood frame construction at some date over the years.

The ground floor space at the rear of the structure behind the vision center was vacant. It had been leased out for a variety of retail and office tenants over the years, but was vacant for at least a year before the fire. The only access to the area under the vision center was through this space, through a number of rooms and doorways. There was no stairway connecting the vision center with the level below.

The vision center extended into part of the upper level of the rear structure and used part of this space as a lab. There may have been an additional section at the rear of the upper level that was unused, with access from the vacant office space.

Special Risk Factors

The arrangement of the interconnected buildings created some very unusual and dangerous conditions for firefighters. From the vision center there was no access from the upper level to the lower level, except to go around to the rear of the building and enter from Crom Street, through the vacant office space. This also created dead end spaces on both levels, estimated at more than 140 feet from the street entrances on each level, where the only way out was the way a firefighter would have entered.

From the stationery store the only rear exit was the stairway down to the pizza parlor and out to Crom Street, which was a long and difficult path. Access to the basement storage level below the stationery store was available, but also very limited.

There were no openings for people to pass from the occupancies on one side of the center dividing wall to the other, but there were numerous openings where smoke or flames could extend through these walls. The false mansard front that had been built over the store fronts was an additional path for smoke travel or fire extension.

Finally, the aged condition of the buildings would have been a major concern. The wood joists were in questionable condition after more than 100 years in place and an unknown number of events, including at least one major fire that destroyed the upper floors.

While these occupancies were only two blocks from the fire station and most of the residents of Pittston had been in and out of them for decades, the fire department did not have a pre-fire plan of the buildings and none of the members reported having an intimate knowledge of the interior arrangement or construction details.

Exposures

Exposures were not a major problem at this incident. The fire buildings were located between a two story bank building, estimated to be 40 years old, and a newer single story YMCA building. The exterior walls of both exposures were brick and concrete construction, abutting the exterior brick walls of the fire buildings. Both exposures had windowless walls, taller than the fire building.

At the rear of the bank was an open parking lot. The front and rear exposures were streets, with single story occupancies across Main Street and a parking lot across Crom Street.

FIRE ORIGIN

The fire was determined to have originated in the vacant area under the vision center from a fault in an electrical conduit. The power supply for the vision center was run from the rear of the building to a panel on the ground floor at the front. The wires were enclosed in conduit that was attached to the underside of the wood joists supporting the ground floor. Although the power had been shut off to the vacant part of the building, this line was still energized to supply power for the occupancy above.

Due to the very cold weather over the weekend, the electrical heaters may have run continuously, causing an unusual current draw through the wires. The continuous current flow would cause the wires to overheat, particularly in an area where the conduit may have been damaged, even if the current was insufficient to blow a fuse or trip a circuit breaker. The overheating is believed to have been sufficient temperature to cause a smoldering ignition of one of the floor joists, approximately 60 feet back from the front of the structure.

The fire may have smoldered undetected for hours. The vacant area had an opening to an old chimney flue or vent stack in the exterior wall, which may have allowed the smoke to escape during the early stages of the fire. The downtown area was sparsely populated due to the snowfall that had started on the previous day, and even if someone had seen smoke coming from the stack it would have looked like smoke coming from a chimney.

The pizza parlor was open until 11 p. m. on the night of the fire. The owner reported that he left the building around 11: 30 p. m. and noted nothing unusual.

A snow removal worker noticed smoke coming from the false front of the stores on Main Street at approximately midnight on March 15, 1993. He called in by radio to the public works dispatcher who turned in the alarm.

Response

The alarm was transmitted by the Pittston Police dispatcher at approximately midnight[2] and the two pumpers immediately responded with the career driver and one volunteer firefighter on each

---

[2] The times of the dispatch, arrival and subsequent events are estimated from witness accounts. There was no recording of the radio traffic or other specific time reference to establish a more accurate time sequence.

vehicle. The other volunteers were alerted by radio and responded directly to the scene. Captain Lombardo, who lived only a few hundred feet from the scene, arrived at almost the same time as the first pumper, which had only a two block response.

Arriving at the scene they noted a moderate amount of lazy gray smoke coming from the eaves over the storefronts, suggesting a minor interior fire of some type. One pumper stopped at the front of the building, while the second pumper laid a supply line to it from the hydrant at Main and Water Streets. A 1-3/4-inch attack line was pulled as the arriving members dressed and prepared for entry.

The fire chief, who was at his residence, was unable to extricate his private vehicle from the snow to respond on the call. Another member of the department who was responding to the alarm picked him up and transported him to the scene. As they arrived they passed behind the fire buildings and noted no evidence of fire or smoke. The first indication of a fire noted by the chief was the smoke coming from the false front over the storefronts.

Noting that there was a possibility of a working fire, the chief instructed the dispatcher to request mutual aid from West Pittston. The West Pittston Volunteer Company responded to the scene from their station just a few blocks across the river. The West Pittston engine company laid a supply line to the front of the buildings from the opposite end of the block, while the ladder truck stopped at the rear on Crom Street (see diagram on following page).

Initial Entry

The initial entry was made into the stationery store, since it appeared to be smoke filled through the front windows. Forcible entry was made through the front door and glass was removed from the front windows, then the line was extended into the store, searching for the fire. The team, which included Captain Lombardo, wore full protective clothing and used self-contained breathing apparatus. They worked their way through the store, all the way to the rear, without encountering any sign of heat or fire. The line was then extended down the basement stairs, into the pizza parlor, and eventually all the way to the door on Crom Street. The interior team had no portable radio to report their progress back to the fire chief, but at the doorway they made contact with firefighters who had gone around to that side and reported that they could not find any sign of the fire.

The line was withdrawn back out to the front of the building where the first entry team had to change the cylinders on their breathing apparatus. Approximately a dozen Pittston firefighters were on the scene, along with a similar number from West Pittston. The amount of smoke coming from the buildings had increased, but still suggested a relatively minor interior fire. The fire chief believed that they would eventually locate a burning piece of furniture or some other easily controllable fire somewhere inside.

The chief had directed one of the assistant chiefs to return to the station and to bring the aerial platform to the scene of the fire. The aerial platform vehicle was positioned in front of the fire buildings where it could be used to supply power for portable lights and fans. A ground ladder was raised and a crew went to the roof to evaluate the need for vertical ventilation. By this time, approximately thirty minutes after the initial alarm had been transmitted, the personnel on the West Pittston ladder truck had noted heavy smoke coming from the side of the building, near the point where the side wall intersected with the wall of the bank building. This is close to the area where the old chimney flue was located.

Second Entry

The determination was made that the fire must be in the vision center side of the buildings. The front windows of this occupancy were removed and forcible entry was made through the front door. Although it was also smoke filled, the smoke was not alarmingly heavy and the line was again extended inside by entry teams wearing full protective clothing and self-contained breathing apparatus.

The entry crews had some trouble navigating through the smoke filled office, but still reported finding no indications of the seat of the fire. One team used up their air supply and came outside. They were replaced by Lombardo and Insalaco, who had responded from West Pittston on the mutual aid request. Taking over the line they continued to search for the fire.

A second crew donned SCBA and followed the line into the building to back-up Lombardo and Insalaco. They reached the first team, but one of the team members of the second team was inexperienced, which caused him to become anxious working in the smoke filled interior. His partner escorted him back out to the front of the building, where they reported to the fire chief that Lombardo and Insalaco appeared to have located the fire in an interior room. This was estimated to be nearly an hour into the incident.

Another two member entry team was assembled and the members followed the line back where Lombardo and Insalaco were last seen. As they worked their way back they encountered much greater heat and came upon an area where flames were coming up through a large hole in the floor. The hoseline appeared to extend into the crater and there was no sign of Lombardo or Insalaco. They quickly returned to the exterior to report their findings.

Rescue Attempted

By the time they reached the street it was obvious that fire conditions were changing rapidly. The smoke coming from the front of the building was hotter heavier and the crew on the roof reported that the heat and smoke issuing from their vent hole had increased rapidly. A second attack line was advanced into the building, but the crew could not reach the area where the floor had collapsed. The fire was rapidly involving the ground floor area and no access to the basement could be located.

One of the assistant chiefs took another crew around to the Crom Street side of the buildings and forced entry through the door into the vacant office space. A hoseline was taken from the second Pittston engine through this door and extended back through the offices toward the front section of the building. Initially only light smoke was encountered, but as they reached deeper into the building they encountered heat and heavy smoke that stopped further penetration. They were unable to reach the area under the front section of the building before they were forced to retreat from the building.

Defensive Operations

Fire was rapidly spreading through the vision center on the upper level and through the spaces below, and soon flames were visible in the stationery store. The situation became a defensive operation as the fire extended throughout the interconnected buildings. The aerial platform was set up in the front, and the West Pittston aerial ladder was set up in the rear parking lot to apply elevated master streams to protect the exposures. Additional mutual aid companies responded, but they were unable to prevent the total involvement and destruction of all four structures. The fire was confined to the complex of four structures and was brought under control by mid-morning.

Body Recovery

It was known almost immediately, when the floor collapse was discovered, that firefighters were missing, but the specific number and identities of the missing members was in doubt. There was no formal system for accounting for members on the scene, and the interior crews had rotated at least twice while searching for the fire. The two missing members had responded with two different companies and Insalaco was wearing a turnout coat labeled with the name of a third department and the rank of assistant chief, which added to the confusion. It was only by a process of elimination that the personnel at the scene were able to conclude that Lombardo and Insalaco were missing and presumed to have fallen into the basement.

By the time the fire was brought under control, the roofs, floors, and some of the walls had collapsed and additional sections of the ice encrusted brick walls were in danger of collapse. For most of the morning crews worked to try to find a way to penetrate the mass of rubble to search for the bodies of the missing firefighters. They eventually discovered an abandoned coal bin in front of the vision center, with a small access cover built into the sidewalk.

The Scranton Fire Department's rescue squad responded to the scene and, after a backhoe had been used to provide a larger access into the coal bin, its members were able to drop down and into the front part of the basement storage area. From there they had to tunnel back through the rubble more than sixty feet, handing debris out and passing shoring materials in, before they finally discovered the two bodies. As assumed, the two firefighters had fallen through the floor into the void space and were trapped in the rubble of floor joists and furniture that had fallen through the hole with them. The bodies were carefully removed through the path that had been tunneled in from the coal bin and further exploration of the area confirmed that no additional members had been lost.

ANALYSIS

It appeared from the positions of the bodies and the furniture that had fallen on top of them that most of a room had dropped into the basement without warning. The other firefighters who had seen the fire reported that Lombardo and Insalaco appeared to be fighting a fire that was coming up around the baseboards of a room, well back inside the vision center, when they were last seen. Further investigation of the fire cause found that the probable point of ignition was in the same area where the collapse occurred, under the floor where the two firefighters were working.

The fire probably ignited one of the 3 by 10 wood floor joists and may have smoldered for hours before it was discovered. Large beams of this type have been known to bum for more than 24 hours before open flaming was observed. The electrical short could have ignited more than one joist or the fire may have spread at a slow rate in the very old wood.

The joists may also have been weakened by age and rotting, so that they could have been much weaker than one would expect from their appearance.

When the fire finally grew to a stage that significant amounts of smoke were produced, it was still contained by the solid wood decking over the floor joists. The crews spent an estimated 60 minutes searching for the fire without finding it, or even detecting a level of heat on the upper level that would have been alarming. By the time the fire became visible on the ground floor level, collapse was imminent.

The circumstances suggest that the collapse occurred totally without warning. Some of the personnel outside reported that they heard a loud cracking noise or a "pop" just before the heat and smoke conditions began to change rapidly. Within minutes the appearance of the situation changed from non-threatening to an obvious major fire.

There was no access from the ground level of the vision center to the space below. Although this space had been used by the previous occupant of the vision center for storage, the only way to check this area would have been to enter from the rear street, the way the rescue attempt was made.

Both firefighters were found to have been using their self-contained breathing apparatus at the time of the collapse and were properly attired in full protective clothing. While these items provided as much physical protection as is generally feasible for interior structural operations, it appears that they quickly succumbed to the combination of their fall and entrapment in the fire area. Examination of the personal protective clothing and equipment revealed no deficiencies.

At least one of the firefighters had a PASS device attached to his SCBA. It was impossible to determine from the damaged parts if it had been turned on or operated during the entrapment. No rescuers reported to have heard a PASS device operating.

LESSONS LEARNED

Several points need to be considered with respect to the way this fire presented itself and the actions that were taken by firefighters based on this information.

1. **Command officers must consider the possibility that a fire which cannot be located is attacking the floor below the search teams.**

    One of the important lessons to be learned from this fire is the danger of a fire burning undetected below an area where firefighters are working. Several similar situations in the past have had similar consequences.

2. **Officers must track the passage of time and assume a fire that cannot be located may be a growing threat.**

    The estimated time from arrival to collapse at this incident is one hour. For this entire period firefighters looked for a concealed fire that gave evidence that it was relatively minor. They continued to operate in a "minor fire" mode, despite the prolonged time without locating the source of the smoke. Officers must maintain an accurate awareness of the passage of time and, if the fire cannot be located, the assumption must be made that it is likely to become more serious.

3. **Infrared heat scanning devices can provide valuable assistance in locating hidden fires.**

    A hand held infrared heat scanning device could have proven invaluable at this incident by helping the interior crews to quickly locate the hidden fire below the floor. These devices have been available for years and are reliable and relatively inexpensive.

4. **Old buildings can be death traps.**

    Buildings that are old and have been renovated numerous times are often exceptionally dangerous to firefighters. They may have inaccessible void spaces, unknown paths where fire can build and spread without being detected, and openings where smoke can migrate to confuse firefighters who are looking for the fire's actual location. Regardless of general appearances, they may have major structural weaknesses that have developed over the years.

5. **Pre-fire plans are essential for complex structures.**

    The complex of structures involved in this fire was extremely complicated and contained several features that should have been recognized as both problems and dangers to firefighters. These factors could only have been recognized through pre-fire visits and should have been recorded in a standard pre-fire plan format to support the officer in command of a fire at this location. It is interesting to note that most of the local firefighters were somewhat familiar with the buildings but not aware of the details of construction and arrangement.

6. **Incident management procedures should be practiced and utilized at all fires.**

    The direction of operations at this incident was conducted without the benefit of a standard incident management system or structure. The fire chief did not have the support of established systems to process information, analyze problems, supervise interior operations, communicate with interior crews, or support a complicated interior operation. The lack of a safety officer and the inability to communicate with interior crews were serious problems in this case.

7. **A personnel accountability system should always be used, particularly at structure fires.**

    The establishment of effective accountability systems for all personnel operating at the scene of fires, particularly those working in interior operations, has become a standard safety practice. This type of system can greatly reduce the risk of overlooking personnel when a building must be evacuated. It also reduces the probability that personnel may become trapped or incapacitated and that no one would be aware that they were missing.

Note: *The similarities of this fire should be compared with the incident in Brackenridge, Pennsylvania in 1991 and the 14th Street Collapse in New York City in 1966; two fires that had similar circumstances and lessons learned and which claimed the lives of 16 firefighters.*

APPENDIX A
Photographs

Photo by Ty Dickerson

Void in brick wall (foreground) was a flue from a heating unit (previously removed) that may have allowed smoke from the basement to vent to the outside in the early stages of the fire.

Appendix A (continued)

Photo by Ty Dickerson

The wall of the front portion of the stationery store remains standing, although the interior is destroyed and the roof has collapsed.

Appendix A (continued)

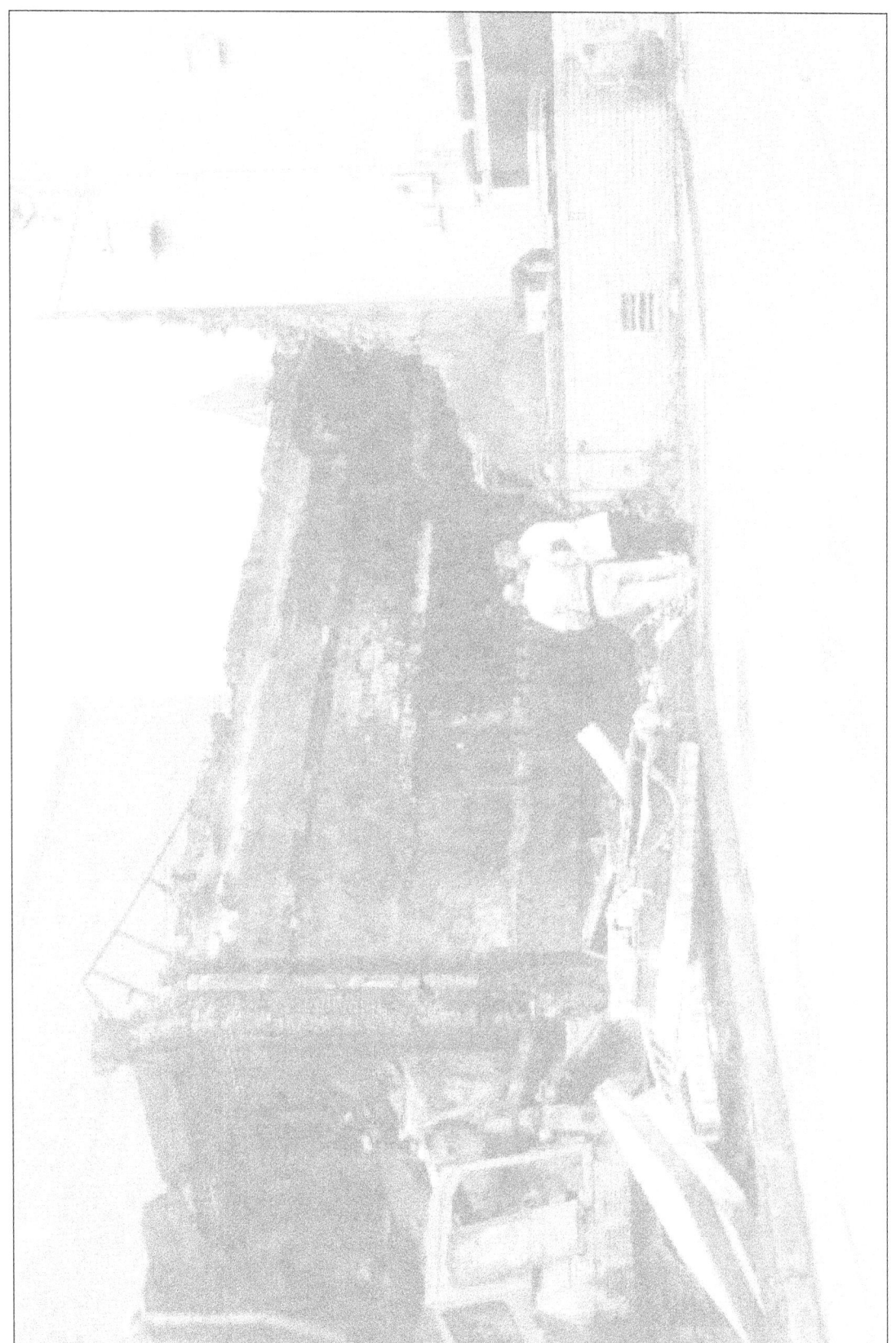

Photo by Ty Dickerson

View of the area of fire origin and the area where the floor collapse occurred, after most of the debris has been removed. Indentations in the brick wall indicate where floor joists were supported.

Appendix A (continued)

Photo by Ty Dickerson

After demolition of the rear part of the fire occupancy, the narrow width of the fire buildings is evident. The demolition was necessary to allow investigators to access the area of fire origin. The boarded-up doorway was the entrance to the pizza restaurant.

Appendix A (continued)

Photo by Ty Dickerson

The outline of the upper floors of the original structures is visible in the wall separating the fire buildings from the exposed bank building (background). The void space above the entrances to the two buildings is also visible. The condition on arrival was smoke coming from this space.

Appendix A (continued)

Photo by Ty Dickerson

Collapsed roof section from the mid-section of the stationery store. The original walls of the front and rear buildings can be seen in this photo. The opening at the lower level is the original alley, which was filled in by construction joining the front and rear buildings. This opening provided an open path for fire extension between occupancies.

Appendix A (continued)

Photo by Ty Dickerson

Several days after the fire, hundreds of floral tributes to the fallen firefighters have been placed on the sidewalk in front of the fire buildings.

Appendix A (continued)

Photo by Luke Alar

Approximately 16 hours after the fire the upper floor of the rear portion of the vision center building has collapsed and heavy equipment has been moved into the bank parking lot to begin debris removal. The door in the single story section that remains standing was the access to the vacant office occupancy. The rear wall of the front section of the building is also visible.

Appendix A (continued)

Photo by Luke Alar

Approximately 16 hours after the fire, a backhoe is in front of the occupancy of fire origin to open an access to the coal bin under the sidewalk and provide access to the basement.

www.ingramcontent.com/pod-product-compliance
Lightning Source LLC
Chambersburg PA
CBHW081247170526
45165CB00009B/3230